Parasites
THE ENEMY WITHIN

by
Hanna Kroeger MsD

Illustrated by
Alberto Kroeger

ISBN 1-883713-07-2

Table of Contents

It is our deepest prayer that this book serves the needs of our people.

One of the greatest
BLESSINGS
God can give us
is the gift
of someone
who cares . . .

Introduction

A parasite is an organism that derives its food, nutrition and shelter by living in or on another organism. In biology the parasitic way of life is very common. There are more organisms in the world that live as parasites than organisms that live otherwise. Humans may be a host to over 100 different types of parasites. Most people seem to think that a parasitic infection is uncommon, especially here in North America. It is a topic most people don't want to discuss. We assume it is a disease of undeveloped countries. In some instances this may be true, especially with the lack of food sanitation. There are many species of parasites that show no socioeconomic boundaries, and may be found in all climates. Nowadays, with air travel and the added mobility of our population, people may become exposed to a greater threat of parasitic infections. Many immigrants harbor parasites from their native countries and Armed Forces personnel returning home have been known to also carry the infections.

This book is designed not to replace your physician, but to act as a step between the patient and the doctor. None of the formulas listed should interfere with the doctor or his medication. Hopefully this book will act as a self-help manual enabling you to understand more about the problem and some natural help that is available to you and that you can initiate on your own.

Thank you, God Bless you and Good Luck,

Hanna

What you have to know about parasites

Once worms or parasites are established in the body, these invaders do four things:

I. Worms can cause physical trauma to the body by the perforation of the intestines, the circulatory system, the lungs, the liver and so on. When chyme is released into the perforated intestines it oozes into the lymph system. Allergies are the first reaction. In other words, worms can make "swiss cheese" out of your organs.

(Chyme is the mixture prepared in the duodenum so the intestine can absorb it.)

II. Worms can also erode, damage or block certain organs. They can lump together and make a ball, a tumor so to speak. They can go into the brain, heart, lungs, and make untold misery for the host.

III. Parasites have to eat, so they rob us of our nutrients. They like to take the best of our vitamins and amino acids and leave the rest to us. Many people become anemic. Drowsiness after meals is another sign that worms are present.

IV. The last and most important way these scavengers cause damage is by poisoning us with their toxic waste. Each worm gives off certain metabolic waste products that our already weakened bodies have trouble disposing of. The poisoning of the host with the parasite's waste is a condition called "verminous intoxication." It can be very serious for the sufferer and it is difficult to diagnose.

An infected individual may feel bloated, tired or hungry, and also have allergies, gas, unclear thinking and generally may feel toxic. Certain parasites have the ability to fool the body of the host into thinking that the worm is a normal part of the body tissue; therefore, the body will not fight the intruder. The host now works twice as hard to remove both its own waste and that of the parasite.

Parasites can be present in any disease. Most doctors will not necessarily treat every infection unless the infection is heavy and the host is showing serious signs of disease. From his standpoint this may be justifiably so because many of the drugs that are used to treat an infection are very strong. They work on the premise of differential toxicity. This means that the drug is hopefully more toxic to the parasite than it would be to the host. Sometimes this margin is slim. Most people with parasitic infections are also usually undernourished, weak, full of either viral, fungal or bacterial infections, and have various types of chemical and metal poisoning. **God has allowed us some help.** When you treat a person naturally for a parasitic infection it is best to treat the whole person usually involving some form of detox program and a nutritional rebuilding program allowing the body to restore balance and health. This, along with an understanding of how the infection was acquired and how to avoid the infection in the future will allow the individual to become more self-reliant in the area of health care.

There are five different groups of parasitic invaders:

I.   Roundworms
II.   The Single Cell Parasites
III.   The Tapeworm Family
IV.   The Flukes Invaders
V.   Spirochetes

Roundworms

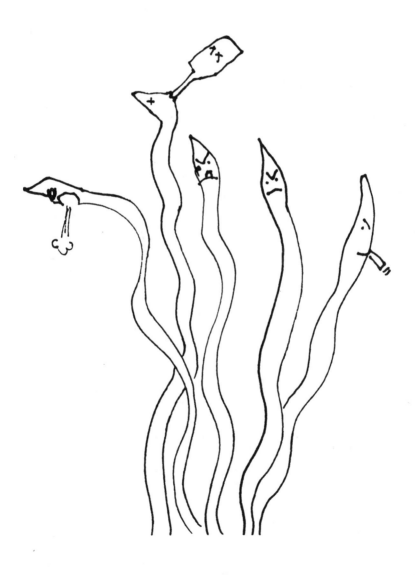

Roundworms have different known species.
By name they are:

Ascara lumbricoides (common round worm)
Hookworm
Strongyloides stercoralis
Ancylastoma caninum
Whipworm
Toxocara canis
Pinworm
Dirofilaria immitis (dog heartworm)
Trichinella spiralis

Symptoms to look for when roundworms are involved:

1). Grinding of teeth at night
2). Intestinal gas
3). Allergies
4). Asthma
5). Snoring
6). Digestive disturbance
7). Anemia
8). Restlessness
9). Weight gain around full moon (sometimes 7-8 pounds)

Roundworm

Roundworms
Ascaris lumbricoides

This parasite is common around the world, especially in warmer climates. Twenty-five percent of the people in the world are thought to be infected. In the southeastern parts of the U.S., up to 64 percent of the people may carry this parasite. It is most common where there are areas of poor sanitation. These worms are large, about the size of a pencil. Humans may be infected with many worms at one time. The infection begins with the ingestion of the eggs which are usually present in contaminated soil, or on fruits and vegetables grown in infected soil. Children are easily infected by eating dirt or putting soiled hands into their mouths. Children are usually more infected than adults. One female worm may release 200,000 eggs per day. Once swallowed the eggs migrate out of the digestive tract into the blood and lymph, passing through various organs including the liver. Eventually the worms will end up in the lungs where they migrate into the trachea and are swallowed and return to the intestines to mature. Occasionally worms may migrate into the eyes, brain, and ears causing extensive damage. In the intestines, the worms consume a large amount of food and give off their waste products. Each worm usually only lives for a couple of years. The worms may release a foreign protein which will cause the host to have an allergic response. Infected individuals may experience abdominal pain, lung infections, eye infections, blood sugar imbalance, weight loss and fatigue.

Roundworms create their own hydrogen peroxide so that their offspring will be plentiful.

For help, try: "Wormwood Combination": Black Walnut Leaves, Wormwood, Quassia, Cloves, Male Fern. 2 caps 3 times daily before meals.

Give someone with worms as much garlic as they can stand, then 2 days later give a laxative. Have them sit in a milk bath sufficient for covering the rectal area. Worms smell the milk and crawl out. Remain in the warm bath for about 1 hour until all the worms are out. This can be rather unpleasant.

Here is an area of the body that often displays signs of parasitic infection, the soles of the feet. Here is a drawing to help:

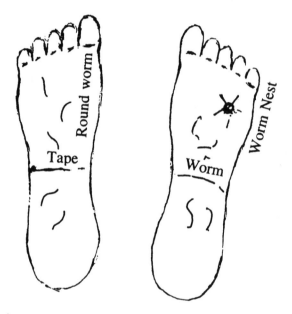

Hookworm

Hookworm
Necator Americanus
Ancylostoma Duodenale

This disease has been the focus of much medical attention. Formerly this has been a major disease in the southeastern U.S. It is most common in warmer, humid regions. It is a disease marked by blood loss. These worms drink large amounts of blood each day. People with this parasite show signs of anemia and malnutrition. During part of the worms' life cycle, these tiny worms must survive as a free living worm outside the body, which lives in the soil or water. This is where humans pick up the infection. We may also get the infection through eating fruits, vegetables or drinking water that is contaminated with this larval stage of the worm. The eggs of the worm are deposited in the soil through unsanitary defecation habits. The eggs then hatch and are free living in the soil. When humans are walking barefooted, or hand gardening, these worms burrow into the skin of the human host. As they burrow into the skin they cause an allergic reaction called ground itch, marked by blisters and itching skin. From here they pass into the circulatory system and eventually make their way to the lung. They then migrate into the trachea and are swallowed. They pass down the digestive tract where they attach to the walls of the intestine and drink blood. Once established, a worm may lay 30,000 eggs a day. The severity of each case is marked by the number of worms in the host's intestine and the nutritional state of the host. Symptoms of hookworm are slow and insidious. People become anemic, malnourished, listless, mentally slow, weak and lazy. They may have abdominal pains with nausea, indigestion, and diarrhea. Hookworms are the only worms with teeth.

For help try:
1. Homeopathic "Hookworm" 15 drops 3 times daily
2. Thyme Tea—2 cups strong thyme tea followed by a dose of Epsom salt. Take Epsom salt half hour after the thyme tea.

3. Old time treatments that may still be of use today:
   1). Male Fern
   2). Eucalyptus
   3). Condurango Bark
   4). Acacia Bark
   5). Oil of Peppermint
4. Also try: Wormwood Combination.

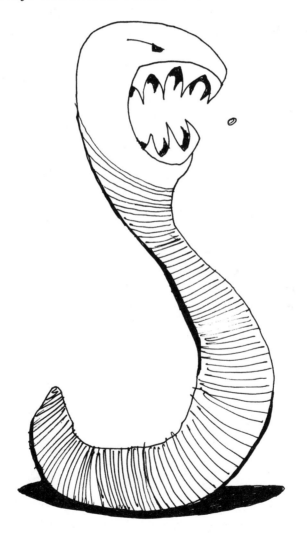

Strongyloides stercoralis

This roundworm is similar to the hookworm. The life cycle begins by a larval worm burrowing into the skin of a human host, passing through the circulatory system and eventually the lungs, up the trachea and is swallowed. It will mature in the intestines. This worm produces young larva in the intestines, where, if given enough time, they will mature and re-infect the original host. This condition is called auto-infection. People with suppressed immune systems can develop heavy infections.

For help try: Treat same as Hookworm

Ancylostoma Caninum

Man occasionally may become infected with a species of hookworm that usually infects other animals, especially dogs. Upon contact with soil contaminated with dog or other animal feces, the larva, which lives in the soil, will burrow into the skin of man. Unable to advance into further tissues, the worms travel around the skin causing a disease called "creeping eruption." Itching and discomfort will occur. After a few months the worm will eventually die.

For help try: Homeopathic Hookworm 15 drops, 3 times daily

Whipworm
Trichuris Trichiura

This is another common parasite, especially in children of warmer, humid climates and in areas of poor sanitation. The worm is small, about 3 to 5 centimeters long. Humans become infected through the ingestion of eggs in contaminated soil or water. Vegetables and fruits may also be contaminated with the eggs. The eggs develop in the small intestine of humans where they later travel into the large intestine to mature. One female may lay 3,000 to 7,000 eggs a day. If humans are infected with only a few worms, the symptoms are less severe. There may be pain in the abdomen, diarrhea, tenesmus, weight loss, nervousness, gas, weakness, insomnia, and "verminous intoxication". Heavy infections show appendicitis, prolapse and edema of the rectum and damage to the intestinal wall.

For help try:
    Homeopathic "Whipworm"
    15 drops, 3 times daily

Sometimes this Homeopathic works well with Homeopathic "Dioxin" 15 drops, 3 times daily

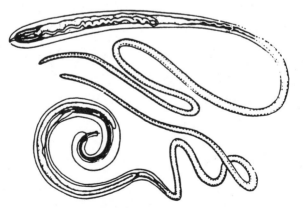

Worms make tumors also

Worms often lump together and make tumors. They are mistaken for fungus tumors.

I have seen this very often, and Professor Brauchle in 1937 warned us to check for worm tumors.

Enclosed is a list of parasites.

> **ALL TYPES OF CANCER CASES**
> are afflicted with worms but
> they do not always make tumors.

To recognize Worm Tumors anywhere in the body, poke around the heels. If sore, watch out.

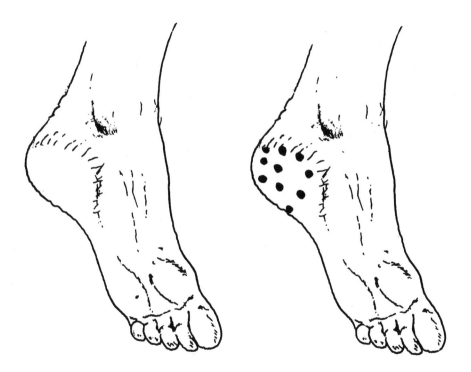

Toxocara Worm
Toxocara Canis   Toxocara Cati

This is the intestinal roundworm of cats and dogs although humans may act as a host. Man becomes infected through ingesting the tiny eggs which are present in soil that has been contaminated by dog or cat feces. Children, due to their playing habits and unsanitary habits around pets, are most commonly infected. Almost all puppies and kittens are infected and humans should be especially careful about handling and cleaning up for these animals. The eggs are ingested through dirty hands, handling pets, or eating fruits or vegetables that may have the eggs on them. The eggs develop in the intestines where they mature and burrow into the circulatory or lymph system, eventually reaching the liver and then the lungs. In animals, these larva complete their life cycle by passing from the lungs into the trachea, being coughed up and reswallowed where they will mature in the intestines. In humans the worm cannot complete the life cycle. Instead it wanders through the body causing damage and a condition called *visceral larva migrans.* The worms may end up in various organs causing a variety of symptoms. They have been known to lodge in the retina <u>where they lead to inflammation and blindness.</u>

Symptoms are:

fever	liver problems
joint pains	lung problems
muscle pains	rash
vomiting	convulsions

For help try:

"Wormwood Combination": Black Walnut leaves, Wormwood, Quassia, Cloves, Male Fern, 2 capsules, 3 times daily, before meals.

When the worms are in the tissue, combine, and take these herbs: Fern (Male), Yellow Dock, Black Walnut, Cloves.

Pinworm

Pinworm
Enterobius vermicularis

This is the most common roundworm in the United
States. It is common to both warm and cold climates and to
all socioeconomic groups. It is most common in crowded areas
like institutions, orphanages, schools and mental hospitals. It
is very contagious and usually many members of one family
are infected. The worms are small from 2 to 13 millimeters,
and white. Adult worms inhabit the cecum and other portions
of the large and small intestines. Female worms crawl down
intestines and pass out of the anus to lay their eggs. A female
may lay 5,000 to 16,000 eggs per day. They may crawl several
inches outside the anus before returning. After a few hours
the eggs are now able to infect other people. The eggs are
light, able to be carried in the air. Sheets, clothes, walls, and
carpets may contain the eggs which stay viable for weeks. A
person may also reinfect himself and other people through
hand to mouth contact after scratching around the anus area.
The most noticeable symptom is the irritation around the
anus and the continued scratching, especially in children.
The worms may also invade the vagina. Children who are infected may show digestive disturbances, nervousness, irri-
tability, and insomnia. Each member of the family may need
to be treated for the infection.

For help try:
1. "Threadworm" Homeopathic 15 drops, 3 times daily
2. Take two cloves of garlic, mash them thoroughly, boil in
6 ounces of milk, let cool and strain. Prepare an enema, inject
4 ounces of this milk into rectum. Do this 3 nights in a row.
Wait 7 days and repeat.
   Cina for children

Dog Heartworm
Dirofilaria immitis

This is a tiny blood and tissue dwelling roundworm that is common in dogs and other mammals and becoming increasingly common in humans. It is found worldwide including the U.S. In man it lives in the heart, blood vessels and the lungs causing a cough and chest pain. It may also cause blockages of the blood vessels, damaging various organs. It is transmitted through the bite of a mosquito. The tiny juveniles (*microfilariae*) live in the gut of the mosquito and are passed from host to host through its bite.

For help try:
Pine Oil
Wintergreen Oil
Mix equal parts, take 3 drops 3x day in apple juice.
"Wormwood Combination": Black Walnut leaves, Wormwood, Quassia, Cloves, Male Fern. 2 caps, 3x day before meals.

Trichinosis

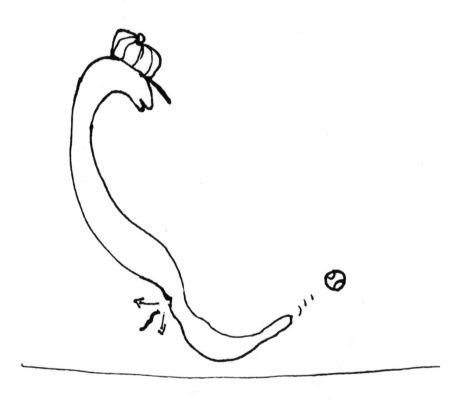

Trichinosis
Trichinella spiralis

This is a tiny roundworm with the ability to cause much disease. This worm has a unique life cycle. It is a disease of carnivores. Any meat eating animal may be infected, including rats, pigs, bears, dogs, and humans. Man becomes infected through eating the larval cyst present in the muscle of an infected animal, usually a pig. From this cyst, the worm matures and mates in the intestinal area, releasing a huge number of young worms. These juveniles now burrow out of the intestines causing diarrhea, intestinal pain, fever, and symptoms of a low grade infection or food poisoning. These worm's waste are toxic to the host, adding to the discomfort. This migration from the intestines into the blood vessels causes a swelling of the tissues, especially around the face and hands. The wandering larva may now also cause symptoms such as pneumonia, encephalitis, brain and eye damage, kidney damage, and many other diseases. The larva will eventually settle into the muscles, where they burrow into the muscle fibers. Now they may be found around the heart, brain, lung, around the face, throat and jaw. The worms stay in the muscles where they form cysts which will eventually calcify and harden. Humans should be especially careful when eating pork, even a small taste of undercooked sausage can lead to a huge infection. Cooking pork in a microwave does not assure that the cysts have been killed. Pork should be cooked until there is no pink left in the meat.

For help try:
3 drops of Oil of Wintergreen in 1 teaspoon of Blackstrap Molasses, twice daily for 3 months. Now after 3 months you must destroy the calcium cysts that were formed. Take Magnesium Oxide tablets: 3 tabs, twice daily for 3 weeks.

Herbs that have anti-parasitic properties

Here is a list of herbs that have been used to help with parasitic infections. Some of the herbs work best in combinations and are sold as capsules. A few of the herbs are very strong, so it is best to get advice from a professional before using. When you are taking herbal combinations for parasites, it is best to take them before meals on an empty stomach. Also, most worms are active during the full moon cycle, it is easiest to treat them during a full moon. Start taking the herbs 5 days before the full moon and continue until you finish the bottle or for about 2 weeks.

Pumpkin seeds
Garlic   Effective as an herbal
Cramp bark   combination
Capsicum   called "Rascal"
Thyme

Black Walnut leaves   .   Effective as an herbal
Wormwood   combination (called
Quassia   "Wormwood Combination")
Cloves   This combination removes
Male Fern   the skin from the worm.

Wormwood 2 parts
Sage 1 part   Effective as a combination
Capsicum 3 parts

Black Walnut
Sassafras   Effective as a combination
Pine needles   for bladder worms

Single-Celled Parasites

Protozoa Infections

Protozoa

This is a large classification of parasites. They are micro-
scopic single-celled animals. More people around the world are
killed or harmed by protozoan infections than by any other
type of parasitic disease. They are essentially found everywhere in our environment. Most are able to form a resting
stage that is very resistant to temperature, chemicals and dry-
ing. This cyst stage is usually the most infectious to man. The
cysts are small and light and easily ingested. Many people
have been exposed to these protozoans yet their immune systems keep the organisms under control, but individuals with a
toxic condition, weakened immune systems or who are under
trauma or stress cannot fight off these parasites. Many of
these parasites are easily passed from person to person during
their cyst stage under crowded, unsanitary conditions. Some
of them are exclusively found in institutions such as orphanages, mental hospitals, day care centers, etc. Also the impor-
tance of the infected food handler should not be overlooked.
Many people may become infected due to the unsanitary
habits of one cyst-passing individual.

Different areas of the body can be affected causing a wide
variety of symptoms. Within this general grouping includes
protozoans of the genera:

Isospora - found in the intestines destroying the surface layer
  of cells.

Pneumocystis - infects the lungs of young children and individuals with suppressed immune systems.

Dientamoeba and *Chilomastix* - infects the digestive tract.

Sarcocystis - infects the muscle tissue while releasing the
  toxin sarcocystin.

Balantidium - invades the intestines and eventually other
  areas while it releases the tissue-destroying enzyme
  "hyaluronidase."

Cryptosporidium and *Babesia* and *Retortamonas* - other less
  common tissue protozoans.

Here is a quote from Dr. Bingham M.D., author of the book *Fight Back Against Arthritis*:

"Our environment contains free-living one-celled parasites on the surface, soil and in fresh water. They form cysts which float in the air and are continually inhaled. These amoeba-type one-celled animals are found in healthy human tissues and exist in large numbers in human cancer tissue, also in tissue taken from cases of rheumatoid diseases. The presence of these protozoa in the body represents the source of constant antigenic stimulation believed responsible for rheumatoid diseases, as well as development of myelomatasis and lymphomata. . . . The blood of healthy people contains antibodies against these organisms. In certain susceptible individuals they may migrate from the gastro-intestinal tract into the blood and then into the joints and other body tissues. Severe stress, injuries and illnesses may precipitate diseases by lowering general tissue resistance which under favorable circumstances might not cause symptoms."

Dr. Bingham wrote the following in his book:

Diseases that may be associated with protozoan infections include:
Arthritis
Asthma
Colitis
Degenerative muscle diseases
Diabetes
Elevated white blood cell count
Hodgkins Disease
Leukemia
Lymphoma
Multiple Sclerosis
Ovarian cysts
Psoriasis
Pyorrhea

Cutaneous ulcers, swellings, sores, papular lesions, and itchy dermatitis can all result from protozoan invasion.

Besides causing many different illnesses these parasites give off toxic waste products which include the compounds: acetate, ammonia, fatty acids, lactate, pyruvate and carbon dioxide.

What can we do? As long as you fight protozoa infection, eliminate the following foods: white potatoes, eggplant, tomatoes and red peppers.

Before each meal, take: 3-cuprum, 6-12x potency; 3-ipecac, 6-12x potency. 10 drops of homeopathic drops 38-6. Drop this in a little water.

If there are marks at the indicated areas on the insole of the foot, it shows protozoa is present.

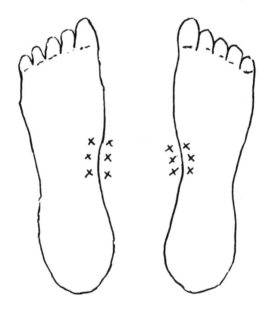

Toxoplasmosis

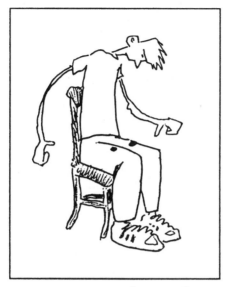

These scavengers make you depressed

Toxoplasma gondii is a crescent shaped intracellular protozoan. It is very common, very contagious and has a worldwide distribution. Many animals act as hosts, including man. Most people have been exposed to this parasite, showing antibodies for it, but have few symptoms. Like most parasitic diseases, the symptoms may wait for years to show up, starting when the body becomes weakened. People with low immune systems are more susceptible. There is a wide range of sometimes vague symptoms. Mild cases show signs, and can be mistaken for a severe case of the flu or infectious mononucleosis with chills, fever, headache, swollen lymph glands, low blood sugar, rash, anemia, swollen spleen and extreme fatigue. As the parasite infects the organs, the symptoms become even worse. The liver, brain, heart and eyes can all be damaged. Toxoplasmosis may show up with:

1) Hodgkins disease

2) leukemia
3) heart problems
4) blindness
5) pneumonia
6) low blood sugar

Pregnant women or women with babies should be the most cautious. Toxoplasmosis can be very serious for the fetus or the newborn. The parasite is able to cross from mother to baby causing brain and spinal cord swelling, eye infection, hydrocephalus (water on the brain), microcephaly (decreased head and brain size), cerebral calcifications, epileptic seizures, mental retardation, spontaneous abortion and still-birth. Humans acquire toxoplasmosis either through eating raw or under-cooked beef, mutton, pork or chicken that have been infected with the organism or by ingesting the cysts after being in contact with contaminated areas such as cat litter boxes, soil, etc. Cockroaches and flies that have been in contact with the cysts then infect food. Almost all house cats have been found to be infected and passing cysts. Pregnant women should be extremely cautious when handling the cat litter box or even the cat. Children playing in sandboxes in which the cat defecates may also become infected. The cysts are extremely small and light and are very resistant to acids and bases but can be killed by heat (thoroughly cooked food) or by low temperatures (hard frozen food).

For help with Toxoplasmosis try:
    1. Oil of sassafras, 2 drops on the soles of feet twice daily; also take "Uplift," a combination of lecithin, licorice root, and anise.
    2. Or try pine oil, thyme oil, equal parts, 3 drops in apple juice 3x day.

Cryptosporidium Parasite

In September of '94, the following story appeared on TV: *City Officials have discovered that a tiny parasite, crypto-sporidium, has become immune to chlorine, and has infested 50% of our drinking water. The eggs are so small that 33% slip through the testing process.* The TV broadcast went on to state that many large cities are infested, specifically mentioning San Francisco, Milwaukee, and New York.

Cryptosporidium creates flu-like symptoms. Some people are affected by stomach cramps and diarrhea. Both the old and young are affected more readily and have the potential for becoming critical much faster than the rest of the population.

Is this silent epidemic becoming a nightmare?

Remember, these tiny critters have become resistant to chlorine, which is outdated in Europe, so it is not surprising that its remedy, The Homeopathic Protozoa Kit, originates from England. The kit includes Cuprum, Ipecac 6x tablets, and Protozoa homeopathic drops. When taken together this combination can help wipe out Cryptosporidium.

Another very good formula is:
Fern
Yellow Dock
Black Walnut
Cloves

Neospora

(The following is an article from *Science News* by J. Raloff.)

"LETHAL LOOK-ALIKE UNMASKED, EXAMINED."

A widespread protozoan infection called toxoplasmosis strikes humans and many other warm-blooded animals, including an estimated 40 percent of all cats. Though many of those infected live symptom-free, others suffer spontaneous abortion, severe illness and even death. . .

Now, a leading toxoplasmosis investigator reports data showing that another, long unrecognized protozoan—able to parasitize many of the same hosts—has for decades masqueraded as a particularly virulent form of the more familiar *Toxoplasma gondii*.

Fourteen months ago, parasitologist Jitender P. Dubey identified and named *Neospora caninum*, isolated from the stored tissues of ten dogs that had succumbed to a virulent toxoplasmosis-like disease. Working at the Agricultural Research Service in Beltsville, Md., he eventually grew *Neospora* in his lab. By infecting laboratory animals with Neospora, he has produced severe toxoplasmosis-like paralysis and death in cats, rats, mice and gerbils over the past year.

Although *Neospora* can infect any tissue, Dubey says, "it is most commonly found in the brain and spinal cord,"as is *Toxoplasma*. The two microbes look similar, except that *Neospora* cysts have a far thicker outer wall.

Most questions about this protozoan's life cycle, prevalence and susceptibility to treatment remain unanswered. "We also don't know whether it is infectious to people," Dubey notes. "But given its similarity to *Toxoplasma*—which infects an estimated 35 percent of the U.S. population—"there is at least a potential for it." *Toxoplasma* can cause central nervous system ailments including paralysis, blindness and retardation.

—*J. Raloff*

Neospora may be a factor in some cases of M.S. For help with this try: Homeopathic "Protozoa Kit," which contains the following: ipecac, cuprum and protozoa.

Sarcocystis

This is a protozoan similar to *Toxoplasma gondii*. It infects muscle tissues causing pain, swelling and degeneration. Sarcocystis releases a toxin called sarcocystin which may affect the central nervous system, heart, lung, adrenal glands, liver and intestines. Just like Toxoplasmosis,we get it through eating under-cooked meat or ingesting spores.

For help try: Same treatment as toxoplasmosis.

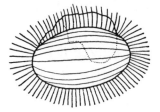

Parasites in the spinal fluid

When a parasite shows up in the spinal fluid there is a homeopathic that will help. People in this situation are nervous, have trouble sleeping, and are irritable.
For help try:
Special X Homeopathic, 3 tablets 2 times daily
Special X Tea, 1 cup 2 times daily.

Trichomonas Vaginalis

This is a species of protozoa that lives in the vagina and urinary tract of females and the prostate, seminal vesicles and urinary tract of males. In females it may cause a burning, itching and chafing of the vagina, a vaginal discharge, frequent and painful urination, and an inflamed urinary bladder. It may also be associated with cervical cancer. In males there may be a small discharge, enlarged prostate, painful urination and inflammation along the urinary tract. It may be transmitted through sexual contact or by contact with contaminated articles. A prolonged infection may cause damage to the reproductive and urinary tissues.

For help try: "Trichomonas" Homeopathic 3 tabs 2x day
Marshmallow root tincture or tea

To repair blocked female tubes rub a few drops of lemon oil over pelvic area twice a day for 3 weeks.

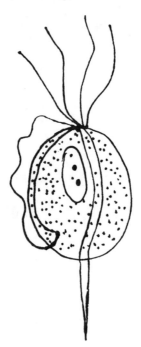

Giardia

Giardia is a single celled, pear-shaped protozoan that lives in the small intestine and sometimes the gall bladder. It is very common and can cause great discomfort and damage to the intestinal walls. Some cases may also show very mild symptoms. Diarrhea is the most common symptom, but mal-absorption, light colored fatty stools, gas, abdominal cramps, lactose and meat intolerance, folic acid and fat soluble vitamin deficiencies may also occur. The severity of the symptoms varies with each case. Individuals become infected either through contaminated food or water containing the cysts or through hand to mouth contact with infected articles like clothes or diapers. Children are more affected than adults. Giardia is very contagious with a large number of cysts being passed in a small amount of feces. All stream and mountain water should be considered infected, because many animals besides humans may act as hosts. The cysts have also been found in municipal water supplies. Chlorination and filtering do not always kill the cysts. All questionable water should be either boiled or treated with iodine.

For help with Giardia try Homeopathic "Giardia",15 drops, 3 times daily.

For full result combine with Goldenseal tincture. 7 drops 3x daily.

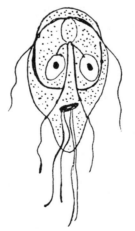

Ameba

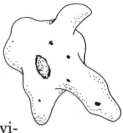

The ameba is a single-celled, micro-
scopic organism with an irregular shape.
It can be found in water, on fruits and
vegetables, in soil and many other damp
environments. It can survive outside ideal
areas by forming a cyst, which can remain vi-
able for long periods of time. Amebiasis is one of the most cos-
mopolitan of the parasitic diseases, it can be found everywhere in the world. It is most common where there is poor
sanitation. When either the cyst or the ameba is ingested it
causes a variety of symptoms including diarrhea, constipation, abdominal pain, gas, weight loss, and chronic fatigue.
The susceptibility of the host plays an important part in how
or if the infection occurs and to what degree. The less severe
symptoms usually go untreated. patients with immune system disorders are very susceptible to infection. Corticosteroid
(cortisone) and other drug therapy can provoke the infection.
Stress also seems to play an important role. Diagnosis can be
difficult. Most infections occur somewhere along the digestive
tract, usually by a species called *Entamoeba histolytica*.
Inside the body the amebas erode the tissues by releasing a
proteolytic enzyme creating small ulcers causing a condition
called amebic dysentery. A lesion may be formed in the intestines known as an ameboma, which on an X-ray looks like
a "napkin ring" around the intestines. Through the ulcers the
ameba may enter the bloodstream and eventually reach other
organs like the liver and brain.

For help with amebic infections try:
Homeopathic "Amoeba"- 15 drops 3x daily.

If the infection is in the liver, combine the Homeopathic
with these herbs:
1) Goldenrod
2) Goldenseal Root
3) Cloves
Combined together this product is called "Livah".

Note: The diet of the host can affect the amebic infection. A high carbohydrate diet has been found to make the infection worse. Also, you may disinfect suspect water by boiling or by adding a small amount of iodine to the water. Chlorination does not kill the cysts.

With the liver involved, the host will show: afternoon fever, weight loss, headache, nausea with pain, liver abcesses, dull headache, unclear thinking.

Two species of amebas may cause fatal brain infections: *Aegleria* and *Acanthamoeba*. Both will cause meningitis-type symptoms starting with a severe headache, loss of smell and taste, eventually leading to a coma and death. Both are acquired through swimming in infected water with the amebas entering the mouth or nose.

One more common species of ameba is found in the mouth area, *Entamoeba gingivalis*. It is found in the area between the teeth and gums and around the jaw bones, eating away at these areas causing a variety of mouth disorders.

For help try:
Mulberry twigs boiled in white grape juice, 4 ounces three times daily. For children with cavities, try the following: Give one tablespoon several times a day, have them hold it in their mouth before swallowing it. Try for one week.

Ameba which is widespread in India is kept under control with Jasmine tea. Jasmine tea is a semi-fermented black tea called "oolong" and jasmine flowers are added. Without this national drink, ameba would have destroyed this country by now.

The Tapeworm Family

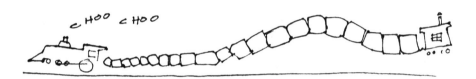

Tapeworms can make
1) Mineral imbalance
2) Thyroid imbalance
3) Intestinal gas
4) High & low blood sugar
5) Bloatedness
6) Jaundice
7) Fluid build-up during full moon

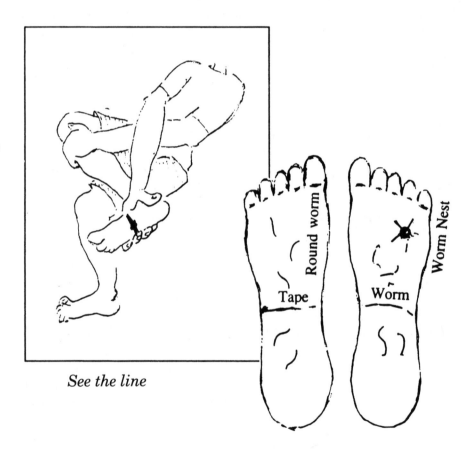

See the line

Tapeworms are long, flat, ribbon-like creatures. They are common in all parts of the world. There are many different species of tapeworms and man is easily a host. Tapeworms do not have a digestive system so they absorb nutrients through their skin. inside humans, tapeworms live in the intestines where they absorb our nutrients, especially vitamin B-12 and folic acid, and give off dangerous waste products. People with tapeworm infections feel toxic, dizzy, have unclear thinking, high and low blood sugar, hunger pains, poor digestion, allergies, are sensitive to touch, and have symptoms associated with pernicious anemia. These individuals have what is called "verminous intoxication," resulting from the worm's wastes. When the worm is rolled up it creates a ball under the right side of the ribs below the liver. Oftentimes tapeworm infections cause a sugar imbalance and people either gain or lose weight. Most tapeworms require an intermediate host besides humans and the worms are named accordingly. In a few instances, humans may serve as both hosts and this can result in a dangerous condition called "cystircercosis" or bladder worms. These juvenile larvae stages of the worms may burrow into various organs of the body.

Beef tapeworm *Taenia saginata*

This tapeworm is found worldwide and is very common in the U.S. It is among the most common types of tapeworms in man. They can become 4 to 8 feet long and cattle serve as their intermediate host. Cattle ingest the eggs which develop into a larval stage inside the muscles. From there humans acquire the infection through eating raw or undercooked beef. Beef that is infected is called measly meat. About 1% of American cattle are infected and only about 80% of the cattle are inspected for this worm. 25% of all infections are missed during inspection.

Bladder worm
Cystircercosis

Man may become infected with the larval stage of certain species of tapeworm. This larval stage is a bladder worm. It can be a large cyst that may lodge in various organs including the heart and the brain. This cyst can be a constant source of irritation causing the body to have an allergic response.

For help try: "Rascal": Pumpkin seeds, Garlic, Crampbark, Capsicum, Thyme.

Or: Black walnut 4 oz. Sassafras 2 oz. Pine needles 2 oz. Make a tea and drink three 8-ounce cups a day.

Bladder worm

Pork Tapeworm
Taenia solium

For this tapeworm the pig is the intermediate host. The dangerous aspect of this tapeworm is the fact that humans may act as both the intermediate host, carrying the larval cyst stage somewhere inside their body, and as the final host carrying the adult worm inside their intestines and passing out eggs, or act as both, a condition known as self-infection. Inside the intestines the adult worm may become 10 feet long. People with adult tapeworm infections show symptoms similar to those discussed earlier. They get the infection by eating raw or under-cooked pork. To become infected with the larval stage (cystercerci, or bladder worm) is easier. Humans ingest the tiny egg through contaminated food, water, soil or hand to mouth contact. This condition may become more serious. Once inside the body, the egg hatches into a larva which will eventually migrate throughout the body to various organs like the muscles, heart, eye, liver, spine and brain. Individuals with this disease may end up with headache, blindness, paralysis, epilepsy, etc.

Broad Fish Tapeworm
Diphyllobothrium latum

This is usually the largest of the human tapeworms, reaching up to 30 feet long. Humans acquire the infection through eating raw or under-cooked fish. Man is the final host for the worm, as it must pass through a tiny copepod then to a fish, and finally to man. Symptoms of infection are similar to other tapeworm infections.

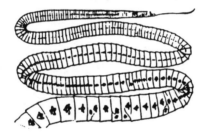

Dog Tapeworm
Dipylidium caninum

This tapeworm is common in dogs and cats everywhere. It is also very common in children due to their close association with these pets. The intermediate host of this worm is the louse or flea. Humans are the final host and they get the infection through ingesting the louse or flea that contains the larva of the tapeworm. Children who are infected show disturbed sleep, grinding of the teeth and intestinal disturbances.

Dwarf Tapeworm
Hymenolepsis nana

Maybe the most common tapeworm is the Dwarf Tapeworm. It is small, only a few centimeters long. This worm needs no intermediate host, so humans, and especially children may become infected through ingestion of the eggs. The eggs may also develop in grain beetles and many other insects and they in turn infect grain. This is how humans usually become infected. Through unhygienic habits, man may reinfect himself with the eggs, which pass out in the feces.

Rat Tapeworm
Hymenolepsis diminuta

This tapeworm is similar to the dwarf tapeworm only larger, up to 1-3 feet long. Rats usually are the final host but occasionally man may be the final host. The intermediate host is the grain beetle or the flour moth. Man becomes infected through eating contaminated insects or grain.

For help for all tapeworms try:
"Rascal":
- Pumpkin seeds
- Garlic
- Crampbark
- Capsicum
- Thyme,

4 caps, 3 times daily.

*Start taking 5 days before the full moon and continue until bottle is finished.

Flukes

It cannot touch me!

Flukes are flat worms that have two ventral suckers that allow them to attach to their hosts.

There are many species of flukes but they are grouped into four types: liver flukes, blood flukes, lung flukes, and intestinal flukes.

Flukes include the genera *Fasciola, Paragonimus, Heterophyes, Schistosoma, Metagonemus, Alaria, Opisthorchis,* and *Dicrocoelium.*

With this many different types of organisms there is a variety of ways to acquire the infection. Humans usually become infected through eating raw or under-cooked fish or crab, eating infected vegetation like water chestnuts, caltop, watercress, or drinking or wading in infected water.

Once inside the body the worms migrate to different areas causing inflammation and damage along the way. They may end up in various organs including the lungs, heart, intestines, brain, urinary bladder, liver, and blood vessels. The worms release many eggs which eventually work their way into either the digestive or urinary tract and are passed out. The eggs may cause extensive damage as they pass through the body. Each egg has tiny spines on the outside that can cause great damage.

The worms also release toxic metabolics that may also damage the host's tissues.

Liver Flukes

Liver flukes are tiny flat worms which undermine the health of the liver. They make holes throughout the liver.

Jaundice
swelling of the liver
general poisoned condition
pain in right side

are symptoms which should be checked for liver flukes disturbance.

Homeopathic vibration "Liver flukes" with "Livah", an herbal combination of Goldenrod, Goldenseal root, and Cloves was found most effective to remove these scavengers.

liver flukes

Lung Flukes

Lung flukes are tiny flat worms which undermine the health of the lungs. Oxygen starvation of the entire blood can be caused by lung flukes. The weakened lungs easily attract other illnesses, such as repeated flu, pneumonia and fungi infections.

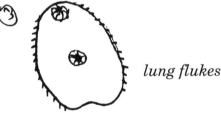

lung flukes

Lung flukes weaken the lungs and perforate the lung tissue. Homeopathic "Lung Flukes" will eliminate these scavengers. Rebuild with natural Vitamin A and "Sound Breath," an herb formula.

Blood Flukes

Blood flukes are tiny flat worms which undermine the health of the blood. They have a hook with which they hook to the blood cells.

Blood flukes cause adults to have blood clots. The heart suffers. Adults wake up in the night several times, and during the day they are sleepy. Eventually affects the bone marrow.

Certain species may also release a proteolytic enzyme that destroys the globin in the blood. Blood flukes are also known to take up certain amino acids, especially argenine, from the host's blood, causing a protein imbalance.

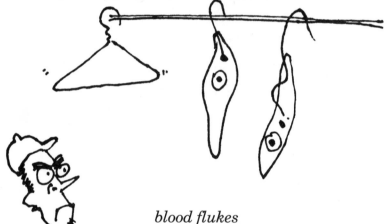

blood flukes

For help try:
Homeopathic "Blood Flukes" 15 drops 3x day
* To rebuild after Blood Flukes take Silica and Collinsonia Root.

FLUKES
Milkweed
Pennyroyal
Black Walnut

This is helpful in Flukes.

Spirochetes

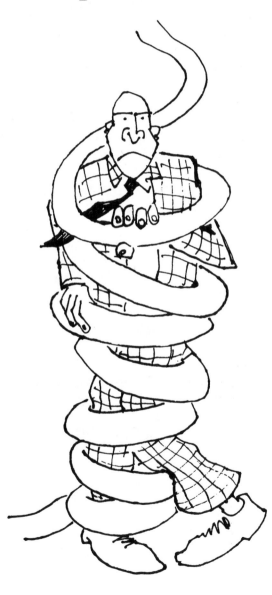

Spirochetes

The development of the electron microscope brought new knowledge and enlightenment to us.

Scientists found a tiny bacterium which is spiral-shaped and called it *spirochete*. In New Zealand and Australia they called this spirochete a parasite which they say comes from sheep and other domestic animals.

Quote from the book, *Introduction to Parasitology* by Asa C. Chandler:

"On the vague unsettled borderline between bacteria and protozoa is a group of organisms, the spirochetes, some of which are waging a frightful war against human life and health."

A spirochete is a spiral-formed organism which does not have the characteristics of a protozoa (not flagellates). It has more characteristics of bacteria.

There are many kinds, the largest being *Spirochaeta*. The genus *Saprospira* and *Cristispira* are smaller, and both are very active in movement. *Treponema* is one of the smallest of the coiled organisms.

Spirochetes are able to bend their bodies, and oscillate while adhering to some object by one end. Spirochetes are responsible for many human diseases such as relapsing fever, infectious jaundice, sores and ulcers. The last two diseases are associated with bacterial organisms.

In Vincent angina, spirochetes are found to be the main cause.

The most important factor coming from the research of scientists is that spirochetes are:

1) widespread

2) difficult to diagnose

3) can attach to Candida Albicans, Epstein Barr, or rundown conditions as in AIDS, cancer, and immune deficient diseases.

Spirochetes multiply in the blood and lymphatic system.

I am not speaking of the species *Treponema*. This is a species of spirochete which causes syphilis. The treatment for this specific spirochete is favorably by drug therapy, not with our herb formula.

Researchers found five different kinds of spirochetes. Wyles disease (a swelling of the liver) is one kind of spirochete. Lymes' disease is another kind.

Another kind of spirochete seems to come from apes and the one I am concerned with makes:

Tiredness
Light fever
Muscle pain
Joint pain
Interrupted sleep pattern
Heavy feeling in legs
Aching pains in feet.

This spirochete attaches easily to people who are:

Run down
Overworked
Immune deficient
After a bout with flu
Epstein Barr

The Herb formula I use contains the following herbs:

Nettle
Yerba Santa
Goldenrod
Organic Tobacco
Monolaurin

The result is most remarkable and amazing. It seems to me that the spirochete is dormant for a long time to break up in situations of stress. Animals respond equally well. Scientists teach that *Treponema* and *Treponema palladium* are spiral-shaped bacteria also. These belong to the physicians' skill and will not respond to "Spirokete," an herb formula.

ALL GONE

SIGNS OF PARASITES IN CHILDREN:

When blisters appear on lower lip (inside mouth) children have worms.

When children wipe their nose and are restless at night is another sign of worms.

If children grind their teeth at night, it is a sign that they have worms.

SIGNS OF PARASITES IN ADULTS:

Chronic Fatigue. Chronic fatigue symptoms include tiredness, flu-like complaints, apathy, depression and impaired concentration.

Immune Dysfunction. Parasites depress immune system functioning by decreasing the secretion of immunoglobulin A (IgA).

Constipation. Some worms, because of their shape and large size, can physically obstruct certain organs, such as: colon, liver and bile duct.

Diarrhea. Certain parasites cause diarrhea.

Gas and Bloating. Some parasites live in the upper small intestine causing both gas and bloating.

Anemia. Worms leach nutrients from the human host. If they are present in large enough numbers, they can create enough blood loss to cause a type of iron deficiency, even pernicious anemia.

Nervousness. Parasitic metabolic wastes can serve as irritants to the central nervous system. Restlessness and anxiety are often the result of systematic parasite infestation.

Allergy. Many allergies are caused by parasites. Some parasites increase the level of eosinophils. If this is the case, eosinophils can inflame body tissue and make allergic reactions to food in particular.

How to
Prevent
Parasites

How to Prevent Parasites

Chief Two Tree, M.D., from North Carolina wrote an interesting article in a leading health magazine. He pointed out that because of the increased pollution and environmental poison, parasites of all kinds are invading our body. Chief Two Tree M.D. said that every day we have to do something to discourage these scavengers and stay free from them.

Chief Two Tree M.D. suggests:
    1 Part Cinnamon oil
    1 Part Tea Tree oil
    1 Part Eucalyptus oil
    Mix with 3 parts Almond oil.
    Take 4 drops once a day.

Worms and scavengers cannot take hold of us if we use apple cider vinegar. One to two teaspoons of it in a glass of water or juice. Or the following mix, which is stronger:
    1 part apple cider vinegar
    2 parts apple juice
    1 part apple brandy
    Take one tablespoon every day in water.

Also helpful is a small handful of mulberry twigs boiled in white grape juice. Take 2 tablespoons of the mixture every day. Or try soaking white figs in water overnight. Next morning add it (figs and water) to your cereal.

Also try the following: rub Sassafrass oil (2 drops) on the sole of each foot 2 times a week. Take 1 drop of oil of Wintergreen with a little honey every other day.

Constipation invites parasites. Lack of gall bladder output invites them also. Make your own salad dressing with apple cider vinegar.

Pumpkin seeds are known to keep worms away. Pumpkin seeds have a good supply of natural L. Tryptophane. Also use garlic as a worm destroyer.

Mix 4 parts ground psyllium seed
1 part Diatomaceous earth (clay)
Take one rounded teaspoon every day for rebuilding and worm destroying.

When an epidemic of any kind appears in Europe the population is advised to moisten the skin (after cleaning it with soap and water) with a little vinegar water:
1 Tablespoon vinegar
3 Tablespoons water

Moisten washcloth and go over the body after cleansing with soap. Soap is alkaline, and vinegar being acid gives you protection against infection and parasitic influences.

A Potion, for Youth Preserved (old formula)

These four, one spoonful each, combine
Juice of apples freshly pressed,
Cider aged past seven days,
Apple-vinegar, tart and brown,
Apple-brandy, clear and strong;
Add to these an ounce of honey,
One scant drop of wintergreen;
Stir them, warm them, mix them well,
And take the tonic every dawn,
Saying this: Now I stay well.

Foods that help the body to prevent parasites:
Here are some dietary suggestions. A diet high in carbohydrates and low in protein has been found to make parasitic infections worse. When the body is in an alkaline condition the parasitic infection sets in. It is best to keep the diet slightly acidic both as a preventive measure and when treating the infection. Foods that help keep the intestines acidic are apple cider vinegar and cranberry juice.

Foods to add to diet
1. Pumpkin seeds
2. Calmyrna Figs - the tiny seeds tear the skin of the worms.
3. Garlic
4. Apple Cider Vinegar
5. Cranberry Juice
6. Pomegranates

Foods to avoid
1. Raw or undercooked beef, pork, fish, and chicken
2. Sugars and carbohydrates
3. Mountain water
4. Water chestnuts and watercress
5. Fruits or vegetables that are unwashed or washed in questionable water

My favorite American Indian, Milo Beaver, showed us that all things are vibrational. A lady in the class checked heavily for the "Wormwood Combination" formula—she admitted that she had had worms for a long time.

Milo Beaver took his holy feather and went over her legs from toe to knee. First back and forth then up and down. First one leg then the other. This lady did not check for the Wormwood formula any longer. Unfortunately it did not work for a person with Giardia. On this example you can see that we have to deal with different vibrational techniques.

By pressing on the point indicated, an enzyme is created which small parasites such as flukes cannot tolerate. Once a day only.

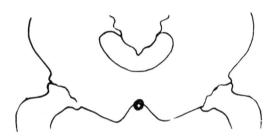

Pelvis

Rebuilding

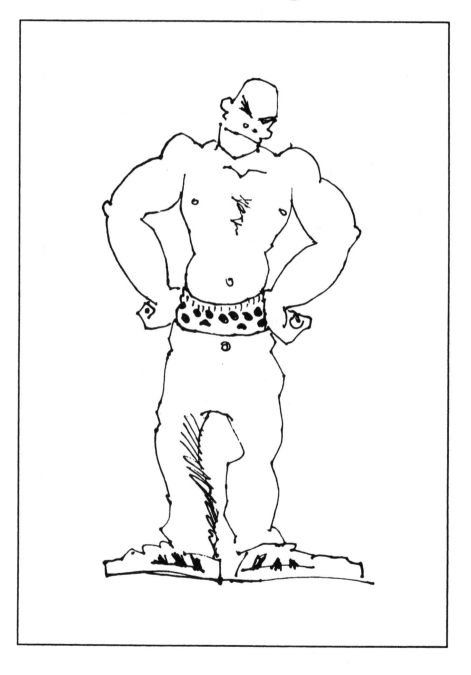

Rebuilding

After you have had any kind of scavengers and after you have cleared your body of them, you have to consider rebuilding your body next. For example, if lung flukes have been in your body, they have made tiny holes throughout the organs, so you have to heal your lungs. Take natural Vitamin A, also E, Sound Breath (an herb formula for the lungs), and use it regularly for 3-4 weeks.

Intestines

If the Parasites have perforated the intestinal tract, the holes also have to be healed. I suggest Acidophilus, Vitamin A, Aloe Vera juice, and Minerals.

Liver

To repair liver damage caused by worms, I suggest: Goldenseal, Goldenrod, Cloves, B-Complex (preferably rice bran), Vitamin E-with maple syrup *and* calcium.
After blood flukes or spirochete, I suggest one gallon Hyssop tea a day for two days (2 gallons total), and no food. This clears the toxic waste the scavengers have left, and heals the damage done.

Protozoa

Two teaspoons liquid chlorophyll in 4 ounces water, 2x daily, Minerals.

Your friends, the Friendly Bacteria

The human body is cut out to manufacture to a great extent its own B vitamins, and certainly will do so when we give it the following chance.

The healthy intestine is loaded with friendly bacteria. Without the help of these workers, assimilation of food would not be possible. People would die of auto-intoxication. A friendly environment is necessary for these little helpers to multiply and work. Milk sugar acts as food for the friendly bacteria and helps to maintain proper acid-alkaline balance.

After a lifesaving dose of penicillin, etc., the unfriendly bacteria are removed, but the friendly bacteria also die. We can easily help this situation by eating lots of yoghurt and using milk sugar on it. Or, a much simpler and faster solution is to drink several bottles of acidophilus milk. To sweeten your intestine and change it to a friendly environment, the following recipe is very good, cheap, and wholesome:

8 oz. powdered buttermilk
8 oz. non-fat milk powder
2 oz. pure fruit pectin
Add 1 quart water and
1 quart of fresh buttermilk
Sweeten each cup with 1 teaspoon of milk sugar.

Take 2 tablespoons of acidophilus milk four times daily and drink one quart or more of the above drink for several days.

Food with preservatives cannot be fully broken down for complete utilization. Lots of friendly bacteria die under the strain. More and more unfriendly bacteria take over, and auto-intoxication begins. Only a healthy intestinal tract can manufacture B complex vitamins.

My best formula—after all these wonderful years of service—is:

Taurin-Dophilus

Here are various herbs that others have had success with.
Please use under supervision of a professional.

American Centuary
Areca Nut
Artemesia Annua
Betel Nut
Betony
Birch
Birds Tongue
Black Walnut Bark
Blue Cohosh
Blue Vervain
Butternut bark or root
    -or- "White Walnut"
Carrot
Cascara Sagrada Bark
Chaparral
Comfrey Leaves
Elecampane
Everlasting
Fennel Seed
Fern-Female
Fraxinella
Garlic

Grapefruit Seed extract
Hedge Garlic
Kousso
Lemon Leaves
Mulberry Bark
Mullein Leaves
Onion
Papaya
Pink Root
Plum
Pomegranate
Pride of China
Pumpkin Seeds
Sage
Senna
Sesame Seed
Tamarind
Turtlebloom
Walnut Hulls
Wafer Ash
Wormseed
Woundwort

Peaceful Meadow Retreat

You are invited . . .

to participate in the upcoming
NATURAL AND VIBRATIONAL HEALING SEMINARS

If you would like to learn how to help yourselves and others with exciting new healing methods, then these weekends are for you.

Learn unusual healing techniques from Hanna Kroeger, well-known lecturer and author of many books.

You will be taught the seven physical and spiritual causes of ill health, how to detect and treat such diseases as Candida, Epstein-Barr and hidden nerve viruses which lower the immune system. Also M.S., Alzheimer's disease and many others.

- Amazing discoveries in Natural Healing
- The best learning vacation you'll ever have

Rev. Rudolf and Hanna Kroeger
7075 Valmont Drive
Boulder, Colorado 80301
(303) 442-2490 or 443-0755

About the Authors

Hanna Kroeger, daughter of a German missionary, was born and raised in Turkey, where she studied natural healing methods under the Oriental and European schools.

She studied nursing at the University of Freiburg, Germany and worked in a hospital for Natural Healing under Professor Brauchle.

Hanna Kroeger is the founder of the Chapel of Miracles and the owner of New Age Foods. Through this little health food store in Boulder, she has been able to help thousands of people improve their health and well-being. She travels extensively throughout the nation lecturing.

She has authored fourteen books, the last of which is entitled *New Dimensions in Healing Yourself*. It is dedicated to the mothers of this nation whom she regards as the torchbearers for a healthier future America.

An annual speaker at NATIONAL HEALTH FEDERATION CONVENTIONS, she electrifies her audience with her profound knowledge and presents her simple and easily available natural help such as food and herbs, to combat many of the plagues of humanity.

Hanna Kroeger MsD is minister of the Chapel of Miracles, Boulder, Colorado.

Books by Hanna

"Wholistic health represents an attitude toward well being which recognizes that we are not just a collection of mechanical parts, but an integrated system which is physical, mental, social and spiritual."

Ageless Remedies from Mother's Kitchen

You will laugh and be amazed at all that you can do in your own pharmacy, the kitchen. These time tested treasures are in an easy to read, cross referenced guide. (94 pages)

Allergy Baking Recipes

Easy and tasty recipes for cookies, cakes, muffins, pancakes, breads and pie crusts. Includes wheat free recipes, egg and milk free recipes (and combinations thereof) and egg and milk substitutes. (34 pages)

Alzheimer's Science and God

This little booklet provides a closer look at this disease and presents Hanna's unique religious perspectives on Alzheimer's disease. (15 pages)

Arteriosclerosis and Herbal Chelation

A booklet containing information on Arteriosclerosis causes, symptoms and herbal remedies. An introduction to the product *Circu Flow*. (17 pages)

Cancer: Traditional and New Concepts

A fascinating and extremely valuable collection of theories, tests, herbal formulas and special information pertaining to many facets of this dreaded disease. (65 pages)

Cookbook for Electro-Chemical Energies

The opening of this book describes basic principles of healthy eating along with some fascinating facts you may not have heard before. The rest of this book is loaded with delicious, healthy recipes. A great value. (106 pages)

God Helps Those That Help Themselves

This work is a beautifully comprehensive description of the seven basic physical causes of disease. It is wholistic information as we need it now. A truly valuable volume. (263 pages)

Good Health Through Special Diets

This book shows detailed outlines of different diets for different needs. Dr. Reidlin, M.D. said, "The road to health goes through the kitchen not through the drug store," and that's what this book is all about. (121 pages)

Hanna's Workshop

A workbook that brings together all of the tools for applying Hanna's testing methods. Designed with 60 templates that enable immediate results.

How to Counteract Environmental Poisons

A wonderful collection of notes and information gleaned from many years of Hanna's teachings. This concise and valuable book discusses many toxic materials in our environment and shows you how to protect yourself from them. It also presents Hanna's insights on how to protect yourself, your family and your community from spiritual dangers. (53 pages)

Instant Herbal Locator

This is the herbal book for the do-it-yourself person. This book is an easy cross referenced guide listing complaints and the herbs that do the job. Very helpful to have on hand. (122 pages)

Instant Vitamin-Mineral Locator

A handy, comprehensive guide to the nutritive values of vitamins and minerals. Used to determine bodily deficiencies of these essential elements and combinations thereof, and what to do about these deficiencies. According to your symptoms, locate your vitamin and mineral needs. A very helpful guide. (55 pages)

New Dimensions in Healing Yourself

The consummate collection of Hanna's teachings. An unequated volume that compliments all of her other books as well as her years of teaching. (155 pages)

Old Time Remedies for Modern Ailments

A collection of natural remedies from Eastern and Western cultures. There are 20 fast cleansing methods and many ways to rebuild your health. A health classic. (115 pages)

Parasites: The Enemy Within

A compilation of years of Hanna's studies with parasites. A rare treasure and one of the efforts to expose the truths that face us every day. (65 pages)

The Pendulum, the Bible and Your Survival

A guide booklet for learning to use a pendulum. Explains various aspects of energy, vibrations and forces. (22 pages)

The Seven Spiritual Causes of Ill Health

This book beautifully reveals how our spiritual and emotional states have a profound effect on our physical well being. It addresses fascinating topics such as Karma, Gratitude, Trauma, Laughter as medicine . . . and so much more. A wonderful volume full of timeless treasures. (142 pages)

Bibliography

Benenson, Abram S., *Control of Communicable Diseases in Man*. American Public Health Association, 1985.

Chandler, Asa C., *Parasitology*. New York, John Wiley and Sons, Inc.

Markell, Edward K. and Marietta Voge, *Medical Parasitology*, W.B. Saunders Company, Philadelphia, PA, 1981.

Schmidt, Gerald D. and Larry S. Roberts, *Foundations of Parasitology* The C.U. Mosby Company, St. Louis, MO, 1977.